Abracadabra! You can add a light and happy touch to your child's learning time! *Homework Magic* will help you reinforce important basic math skills that will build the foundation for future learning. The one-on-one time you spend teaching your child is an irreplaceable gift that will give her or him the extra knowledge that will lead to self-confident, successful school years.

HOW TO USE THIS BOOK

* Choose a place to work with your child that is comfortable, cozy, and relatively free of distraction. This is special time!

* Remove the "magic wand" from the binding of the book. Tear the perforated lines to create the wand. Set the wand aside until it is time for your child to check his or her answers.

* Make sure your child understands the directions before beginning an activity.

* Invite your child to complete the problems on an activity page in pencil, offering as much help and encouragement as your child needs to feel successful.

* When it is time for your child to check the answers, show him or her how to use the wand to reveal the hidden answer. Have fun during this part of your learning time! When your child reveals the correct answer, say, "abracadabra," or "presto," or "wow!" If your child reveals an answer that differs from his or hers... *Hocus-pocus!* Just erase the incorrect answer and rework the problem with your child.

#11089 Homework Magic—Grade 2 Subtraction
rights reserved. Printed in the U.S.A.
pyright © 1998 Frank Schaffer Publications, Inc.
740 Hawthorne Blvd., Torrance, CA 90505

ISBN #0-76820-301-5

ited by Cindy Barden
strated by Nancee McClure
sign & Production by Good Neighbor Press, Inc.

Going to the Dogs!

Fact Families: 1–5

Fill in the missing numbers.

$2 + 2 =$ 4

$1 + 4 =$ 5

$5 - 2 =$ 3

$4 - 2 =$ 2

$4 + 1 =$ 5

$5 - 3 =$ 2

$? + ? =$ 5

$5 - 4 =$ 1

$2 + 3 =$ 5

$? - ? =$ 5

$5 - 1 =$ 4

$3 + 2 =$ 5

$4 + 0 =$ 4

$3 + 0 =$ 3

$1 + 3 =$ 4

$4 - 0 =$ 4

$3 - 0 =$ 3

$3 + 1 =$ 4

$4 - 3 =$ 1

$4 - 1 =$ 3

How can you stop a dog from barking in Colorado?

 Move to arizo"

FS114069 Grade 2 Subtraction

Bear With Us!

Subtracting from 5 or less

Subtract.

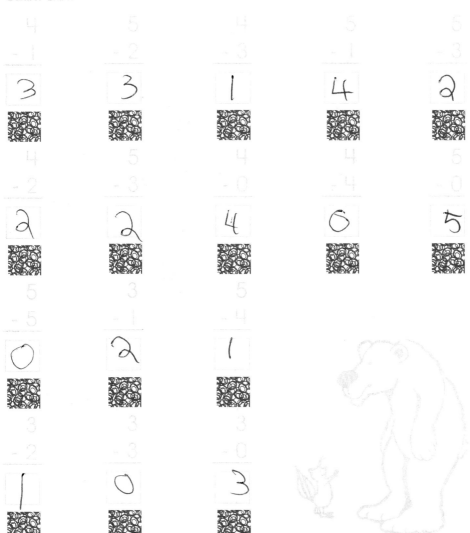

4	5	4	5	5
− 1	− 2	− 3	− 1	− 3
3	**3**	**1**	**4**	**2**

4	5	4	4	5
− 2	− 3	− 0	− 4	− 0
2	**2**	**4**	**0**	**5**

5	3	5
− 5	− 1	− 4
0	**2**	**1**

3	3	3
− 2	− 3	− 0
1	**0**	**3**

How do you get fur from a bear?

FS111089 Grade 2 Subtraction

Rainy Days!

Fact Families: 6-8

Fill in the missing numbers.

$3 + 3 =$ **6** $2 +$ **4** $= 6$ $5 + 3 =$ **8**

$6 - 3 =$ **3** $6 - 4 =$ **2** $3 +$ **5** $= 8$

$4 + 4 =$ **8** $6 -$ **2** $= 4$ $8 - 5 =$ **3**

$8 - 4 =$ **4** $3 +$ **4** $= 7$ $8 -$ **5** $= 3$

$4 + 2 =$ **6** $7 - 4 =$ **3** $2 +$ **5** $= 7$

$4 + 3 =$ **7** $7 -$ **3** $= 4$ $7 - 5 =$ **2**

$5 + 2 =$ **7** $6 + 2 =$ **8** $7 -$ **2** $= 5$

                        $2 +$ **6** $= 8$ $7 + 1 =$ **8**

                        $8 - 6 =$ **2** $1 +$ **7** $= 8$

                        $8 -$ **2** $= 6$ $8 - 7 =$ **1**

Why do mother kangaroos
hate rainy days?

Egg-cellent!

Subtracting from 6, 7, & 8

Subtract.

6 - 2	7 - 3	6 - 1	7 - 4	8 - 6
4	4	5	3	3

8 - 3	6 - 4	7 - 2	8 - 1	6 - 5
5	2	5	7	1

8 - 4	7 - 7	6 - 3	8 - 2	7 - 6
4	0	3	6	1

8 - 1	7 - 5
7	2

Why don't elephants lay eggs?

Frank Schaffer Publications, Inc.　　　　　5　　　　　FS111089 Grade 2 Subtraction

Subtraction Wheels!

Subtracting from 5, 7, & 8

Subtract to complete the wheels.

Why is a pig a tattle tale?

Hip, Hip, Hippo!

Subtracting from 8 or less

Subtract.

$8 - 4 =$ $7 - 7 =$ $7 - 6 =$

$4 - 2 =$ $6 - 5 =$ $8 - 5 =$

$5 - 2 =$ $6 - 4 =$ $4 - 3 =$

$5 - 3 =$ $8 - 6 =$ $7 - 5 =$

$8 - 4 =$

What's as big as a hippo but doesn't weigh an ounce?

Frank Schaffer Publications, Inc.   FS111082 Grade 2 Subtraction

These Make Cents!

Subtracting money

Subtract.

8¢	5¢	4¢	8¢	7¢
− 1¢	− 4¢	− 2¢	− 6¢	− 3¢

6¢	7¢	5¢	8¢	6¢
− 2¢	− 4¢	− 3¢	− 7¢	− 4¢

Don't forget the ¢ sign!

7¢ − 5¢ =

8¢ − 5¢ =

6¢ − 3¢ =

7¢ − 6¢ =

Why does a dog wag his tail?

FS11089 Grade 2 Subtraction

Nine, Ten, a Kangaroo!

Fact families: 9 & 10

Fill in the missing numbers.

$5 + 5 = \boxed{}$ $6 + 3 = \boxed{}$ $7 + 2 = \boxed{}$

$10 - \boxed{} = 5$ $3 + \boxed{} = 9$ $2 + \boxed{} = 9$

$5 + 4 = \boxed{}$ $9 - 6 = \boxed{}$ $9 - 7 = \boxed{}$

$4 + \boxed{} = 9$ $9 - \boxed{} = 6$ $9 - \boxed{} = 7$

$9 - 5 = \boxed{}$ $6 + 4 = \boxed{}$ $7 + 3 = \boxed{}$

$9 - \boxed{} = 5$ $4 + \boxed{} = 10$ $3 + \boxed{} = 10$

$8 + 1 = \boxed{}$ $10 - 6 = \boxed{}$ $10 - 7 = \boxed{}$

$1 + \boxed{} = 9$ $10 - \boxed{} = 6$ $10 - \boxed{} = 7$

$9 - 8 = \boxed{}$

What do you get when
you cross an elephant
with a kangaroo?

Frank Schaffer Publications, Inc.   FS111089 Grade 2 Subtraction

Flying Free!

Subtracting from 9 & 10

Subtract.

9	10	9	10	9
− 6	− 7	− 9	− 5	− 7

10	10	10	10	10
− 6	− 2	− 8	− 4	− 1

9	9	9	9	9
− 4	− 8	− 3	− 1	− 5

Why can't sea
gulls fly over
the bay?

FS111089 Grade 2 Subtraction

Subtraction Wheels

Subtracting from 8, 9, & 10

Subtract to complete the wheels.

Wheel 1: 10−
6, 4, 9, 3, 7, 8, 10, 5, 6

Wheel 2: 9−
9, 2, 5, 6, 3, 8, 4, 7

Wheel 3: 8−
2, 4, 8, 3, 1, 6, 7, 5

Where do sheep get their hair cut?

Number Sentences!

Subtracting from 9 or less

Write the number sentence and answer for each picture.

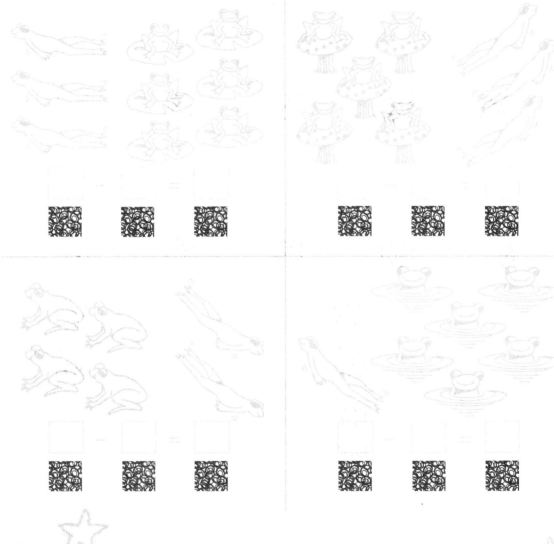

What do you get when you cross a frog and a chair?

© Frank Schaffer Publications, Inc.

FS111089 Grade 2 Subtraction

Money Math!

Subtracting money

Subtract. Don't forget the cents sign.

10¢	8¢	7¢	9¢	6¢
- 5¢	- 4¢	- 6¢	- 4¢	- 3¢

5¢	9¢	6¢	8¢	7¢
- 3¢	- 7¢	- 2¢	- 3¢	- 5¢

10¢ - ☐ = 4¢   7¢ - ☐ = 2¢

9¢ - ☐ = 6¢   8¢ - ☐ = 6¢

9¢ - 4¢ =

10¢ - 7¢ =

How long should a giraffe's legs be?

FS111089 Grade 2 Subtraction

Quiet, Please!

Fact families: 11 & 12

Fill in the missing numbers.

6 + 6 = ☐

12 - ☐ = 6

5 + 6 = ☐   8 + 3 = ☐   7 + 4 = ☐

6 + ☐ = 11   3 + ☐ = 11   4 + ☐ = 11

11 - 6 = ☐   11 - 3 = ☐   11 - 4 = ☐

11 - ☐ = 6   11 - ☐ = 3   11 - ☐ = 4

9 + 2 = ☐   7 + 5 = ☐   9 + 3 = ☐

2 + ☐ = 11   5 + ☐ = 12   3 + ☐ = 12

11 - 2 = ☐   12 - 7 = ☐   12 - 3 = ☐

11 - ☐ = 2   12 - ☐ = 7   12 - ☐ = 3

Why couldn't
the pony talk?

FS11089 Grade 2 Subtraction

The Ayes Have It!

Subtracting from: 11 & 12

Subtract.

11	12	11	12	11
− 7	− 6	− 8	−10	− 6

12	12	12	12	12
− 8	− 9	− 5	− 7	− 3

		11	11	11
		− 3	− 2	− 5

			11	12
			−10	−11

Where do polar bears vote?

15

FS111089 Grade 2 Subtraction

Subtraction Wheels!

Subtracting from 10, 11, & 12

Subtract to complete the wheels.

Wheel 1 (center: 11-): outer ring values 7, 11, 10, 9, 0, 6, 8, 4, 5

Wheel 2 (center: 12-): outer ring values 11, 7, 9, 8, 5, 6, 12, 10

Wheel 3 (center: 10-): outer ring values 10, 4, 3, 7, 5, 8, 9, 6

Why don't mice
like turnips?

Mole Money!

Subtracting money

Subtract.

12¢	9¢	11¢	10¢	8¢
- 9¢	- 3¢	- 8¢	- 5¢	- 5¢

11¢	8¢	12¢	9¢	10¢
- 9¢	- 4¢	- 7¢	- 4¢	- 7¢

7¢	6¢	12¢	5¢	9¢
- 4¢	- 3¢	- 8¢	- 3¢	- 6¢

12¢	11¢
- 6¢	- 7¢

How can you stop moles from digging in your garden?

 FS111089 Grade 2 Subtraction

A Dozen or Less!

Subtracting from 12 or less

Subtract.

12 - 8 = ☐   10 - 7 = ☐

11 - 4 = ☐   9 - 6 = ☐   8 - 3 = ☐

11 - 7 = ☐   12 - 6 = ☐   12 - 5 = ☐

10 - 8 = ☐   9 - 7 = ☐   11 - 8 = ☐

8 - 5 = ☐   10 - 4 = ☐   12 - 4 = ☐

12 - 7 = ☐

10 - 5 = ☐

11 - 9 = ☐

What animals do you always have with you?

Fish Facts!

Fact families: 13 & 14

Fill in the missing numbers.

7 + 7 = ☐   14 - ☐ = 7

7 + 6 = ☐   8 + 5 = ☐   8 + 6 = ☐

6 + ☐ = 13   5 + ☐ = 13   6 + ☐ = 14

13 - ☐ = 6   13 - 5 = ☐   14 - 8 = ☐

13 - ☐ = 7   13 - ☐ = 5   14 - ☐ = 8

9 + 4 = ☐   9 + 5 = ☐

4 + ☐ = 13   5 + ☐ = 14

13 - 4 = ☐   14 - 5 = ☐

13 - ☐ = 4   14 - ☐ = 5

 What kind of fish can you find in a bird cage?

   FS111089 Grade 2 Subtraction

Do Bananas Split?

Subtracting from 13 & 14

Subtract..

13	13	14	14	14
− 8	− 4	− 8	− 4	− 5

13	13	13	14	14
− 2	− 5	− 7	−11	−12

13	13	13	13	14
− 6	−10	− 3	− 9	− 3

14	14
−10	− 6

How can you tell an elephant from a banana?

 FS111089 Grade 2 Subtraction

Subtraction Wheels!

Subtracting from 13 & 14

Subtract to complete the wheels.

Wheel 1 — 13−
13, 2, 5, 11, 9, 3, 10, 7

11

Wheel 2 — 13−
12, 7, 8, 4, 5, 9, 3, 6

Wheel 3 — 14−
11, 5, 3, 9, 8, 7, 14

What do people
in England call
baby cats?

FS111089 Grade 2 Subtraction

Funny Bunnies!

Subtracting from 14 or less

Write the number sentences and the answers.

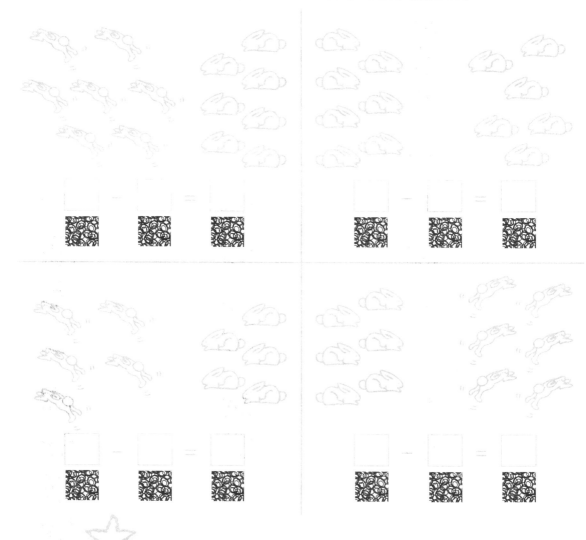

What two-word rhyme describes how a rabbit pays his bills?

FS111089 Grade 2 Subtraction

$8 + 8 =$ ☐

$16 - 8 =$ ☐

$9 + 9 =$ ☐

$18 - 9 =$ ☐

$7 + 8 =$ ☐ $9 + 6 =$ ☐ $9 + 7 =$ ☐

$8 + $ ☐ $= 15$ $6 + $ ☐ $= 15$ $7 + $ ☐ $= 16$

$15 - $ ☐ $= 7$ $15 - 6 =$ ☐ $16 - 7 =$ ☐

$15 - $ ☐ $= 8$ $15 - $ ☐ $= 6$ $16 - $ ☐ $= 7$

$18 - 6 =$ ☐ $9 + 8 =$ ☐ $10 + 8 =$ ☐

$15 - $ ☐ $= 6$ $8 + $ ☐ $= 17$ $8 + $ ☐ $= 18$

$18 - 8 =$ ☐

$18 - $ ☐ $= 8$

What is the best way to keep a skunk from smelling?

Sneaky Snakes!

Subtracting from 18 or less

Subtract.

18 − 6	17 − 9	15 − 8	16 − 9	18 − 8
16 − 8	18 − 9	17 − 4	15 − 7	18 − 9
14 − 9	17 − 8	18 − 7	15 − 6	16 − 7
18 − 4	18 − 5			

What did the cobra say
to the flute player?

FS111089 Grade 2 Subtraction

Subtraction Wheels!

Subtracting from 16, 17, & 18

Subtract to complete the wheels.

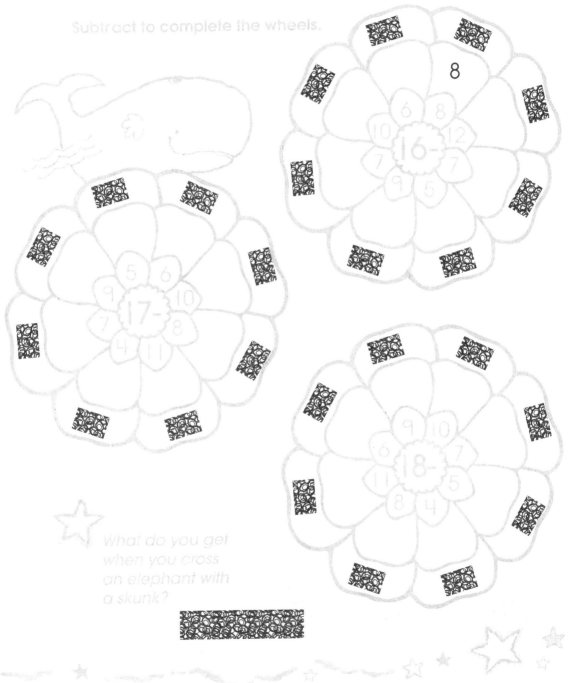

What do you get
when you cross
an elephant with
a skunk?

FS111089 Grade 2 Subtraction

Watch These Carefully!

Subtracting from 18 or less

Subtract.

18 - 9 = ☐   16 - 6 = ☐   17 - 8 = ☐

11 - 8 = ☐   16 - 8 = ☐   13 - 6 = ☐

15 - 7 = ☐   12 - 6 = ☐   17 - 7 = ☐

18 - 7 = ☐   16 - 9 = ☐   17 - 9 = ☐

18 - 6 = ☐   15 - 9 = ☐   17 - 12 = ☐

18 - 8 = ☐   18 - 5 = ☐   17 - 6 = ☐

 Why did the dog
turn around and
around?

FS111089 Grade 2 Subtraction

Number Sentences!

Subtracting from 16 or less

Write the number sentence and answer for each picture.

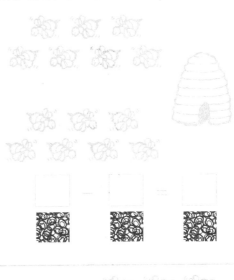

 Why do bees have sticky hair?

FS111089 Grade 2 Subtraction

Batter Up!

Subtracting one-digit numbers from two-digit numbers

Subtract.

57	68	72	89	96
- 5	- 4	- 1	- 5	- 6

55	66	77	88	99
- 3	- 2	- 5	- 4	- 3

28	36	47	26	17
- 5	- 4	- 3	- 4	- 3

			19	18
			- 6	- 5

What do you get when you cross a bat with a mummy?

FS111089 Grade 2 Subtraction

Hungry Cats!

Subtracting two-digit numbers

Subtract.

56	47	65	78	94
− 31	− 23	− 11	− 30	− 32

85	87	75	49	58
− 64	− 17	− 31	− 21	− 22

67	76	85	94	85
− 43	− 54	− 71	− 84	− 32

		76	67	87
		− 53	− 44	− 34

What does a cat make for dinner when it's in a hurry?

Frank Schaffer Publications, Inc.

FS111989 Grade 2 Subtraction

What's for Dessert?

Round numbers

Subtract.

90	80	60	50	30
− 70	− 60	− 60	− 20	− 10

80	40	20	70	90
− 70	− 30	− 10	− 20	− 20

30	50	10	60	80
− 20	− 10	− 10	− 30	− 40

90
− 60

How does a gorilla
make a banana split?

FS111099 Grade 2 Subtraction

Here Comes the Elephant!

Subtracting money

Subtract. Don't forget the cents sign.

18¢	15¢	12¢	11¢	14¢
- 9¢	- 9¢	- 6¢	- 9¢	- 9¢

13¢	16¢	17¢	11¢	17¢
- 6¢	- 8¢	- 9¢	- 8¢	- 8¢

11¢	16¢
- 7¢	- 9¢

13¢	18¢
- 5¢	- 8¢

What do you get when an elephant
walks through a potato field?

Smile!

Fact families: 19 & 20

Fill in the missing numbers.

$10 + 10 = \boxed{}$

$20 - \boxed{} = 10$

$9 + 10 = \boxed{}$ $8 + 11 = \boxed{}$ $12 + 8 = \boxed{}$

$10 + \boxed{} = 19$ $11 + \boxed{} = 19$ $8 + \boxed{} = 20$

$19 - 10 = \boxed{}$ $19 - 11 = \boxed{}$ $20 - 8 = \boxed{}$

$19 - \boxed{} = 10$ $19 - \boxed{} = 11$ $20 - \boxed{} = 8$

$7 + 12 = \boxed{}$ $13 + 7 = \boxed{}$

$12 + \boxed{} = 19$ $7 + \boxed{} = 20$

$19 - 12 = \boxed{}$ $20 - 7 = \boxed{}$

$19 - \boxed{} = 12$ $20 - \boxed{} = 7$

What should you do with a
blue whale?

20 Frogs on 19 Logs

Subtracting from 19 & 20

Subtract.

20	20
- 10	- 6

20	20	20	20	20
- 9	- 5	- 11	- 4	- 8

19	19	19	19	19
- 7	- 4	- 5	- 3	- 8

20	20	19	19	20
- 7	- 13	- 6	- 10	- 12

 What do you do with a green frog?

Subtraction Wheels!

Subtracting from 19 & 20

Subtract to complete
the wheels.

Why were 1992 and 1996
good years for frogs?

Frog Cents!

Subtracting money

Subtract.

What's a frog's favorite drink?

Moo, Moo, to You!

Subtracting two-digit numbers

Subtract.

27	48
- 22	- 43

97	86	35	56	79
- 91	- 83	- 30	- 51	- 78

64	46	54	68	74
- 62	- 41	- 32	- 31	- 52

94	56	47	93	59
- 21	- 42	- 16	- 43	- 36

Why did the cow cross the road?

FS111089 Grade 2 Subtraction

You Can Do These!

Subtracting three-digit numbers

Subtract.

566 - 331	722 - 511

948 - 834	437 - 316	265 - 155	191 - 180

667 - 367	728 - 514	624 - 524	729 - 522

469 - 366	231 - 120	335 - 224	204 - 103

If your cat ate a lemon, what would she be?

Cash Crash!

Subtracting money

Subtract.

$6.34 - 3.04	$7.28 - 5.15
$.	$.

$4.92 - 3.71	$5.17 - 3.16	$2.76 - 1.60	$5.04 - 4.02
$.	$.	$.	$.

$8.71 - 7.50	$7.04 - 5.03	$3.95 - 2.74	$2.99 - 1.11
$.	$.	$.	$.

$7.91 - 5.40	$6.48 - 4.24	$5.16 - 3.15	$8.88 - 4.48
$.	$.	$.	$.

 Which side of a lion has the most fur?

© Frank Schaffer Publications, Inc. FS111089 Grade 2 Subtractio

Rainy Day!

Subtraction with regrouping

Subtract.

```
  92        81        30
- 59      - 36      - 19
```

```
  54        90        36        75        95
- 27      - 25      - 19      - 27      - 26
```

```
  81        42        53        56        43
- 57      - 28      - 39      - 28      - 19
```

```
  29        36        94        87        60
- 15      - 26      - 25      - 45      - 12
```

What kind of weather
do mice like least?

Swim With Care!

Subtracting with zeros

Subtract.

70	80	90	30	20
- 63	- 54	- 83	- 27	- 14

50	60	70	40	30
- 27	- 48	- 55	- 22	- 11

60	80	70	50	60
- 33	- 44	- 39	- 19	- 29

50	90	70
- 47	- 29	- 26

If a shark is after you what should you feed it?

 FS111089 Grade 2 Subtraction

Duck Sale!

Subtracting money

$$92¢ - 86¢$$

$$42¢ - 36¢$$

27¢	34¢	50¢	72¢	83¢
− 19¢	− 26¢	− 43¢	− 63¢	− 76¢

45¢	92¢	61¢	84¢	65¢
− 37¢	− 88¢	− 59¢	− 77¢	− 56¢

61¢	50¢	42¢	38¢	71¢
− 3¢	− 7¢	− 8¢	− 7¢	− 4¢

What do you get when you put 5 ducks in a box?

No Shoes or Socks!

Subtraction with regrouping

Subtract.

56 - 9	82 - 9	91 - 9	67 - 9
32 - 9	27 - 9	18 - 9	84 - 9	91 - 9
77 - 8	63 - 8	52 - 8	37 - 8	48 - 8
85 - 8	93 - 8	47 - 9		

Why don't bears wear shoes and socks?

FS111089 Grade 2

Give Thanks!

Subtraction with regrouping

Subtract.

```
  795
- 388
```

```
  663
- 217
```

```
  671
- 532
```

```
  512
- 106
```

```
  432
- 113
```

```
  840
- 531
```

```
  711
- 506
```

```
  625
- 419
```

```
  230
- 102
```

```
  357
- 138
```

```
  490
- 215
```

```
  323
- 108
```

```
  560
- 240
```

```
  814
- 307
```

```
  274
- 168
```

What should you be thankful for on Thanksgiving?

Snack Time!

Subtraction with regrouping

Subtract.

590 − 554	281 − 235	372 − 326	982 − 447

874 − 337	893 − 857	782 − 768	791 − 578

470 − 458	653 − 129	984 − 538	392 − 368

790 − 85	925 − 780

What's the difference between an
elephant and a doughnut?

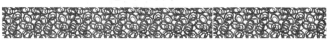

44 FS111089 Grade 2 Subtraction

Froggy Goes A'Counting!

Subtracting four-digit numbers

Subtract.

$$8886 - 4562$$

$$7694 - 3351$$

$$6824 - 5314$$

$$9460 - 5350$$

$$4728 - 3416$$

$$3692 - 1581$$

$$6192 - 4081$$

$$7263 - 1111$$

$$8890 - 6730$$

$$2468 - 1357$$

$$6666 - 3331$$

$$5555 - 2424$$

What kind of frog is the meanest?

Poor Snake!

Subtraction with regrouping

Subtract.

```
  367        456
-   8      -   9
```
[____]   [____]

```
 762       384       971       622       588
-  8      -  7      -  6      -  6      -  9
```
[____]   [____]   [____]   [____]   [____]

```
 391       922       485       915       194
-  3      -  6      -  7      -  8      -  5
```
[____]   [____]   [____]   [____]   [____]

```
 618       427       361       419       288
-  9      -  7      -  3      -  9      -  9
```
[____]   [____]   [____]   [____]   [____]

Why did the baby
snake cry?

Knock, Knock!

Subtract.

83	74	56	49	38
− 62	− 51	− 13	− 37	− 23

$9.55	$8.49	$4.58	$6.19
− 1.15	− 2.25	− 2.25	− 3.11

$8.98	$6.76	$2.85	$4.37
− 4.56	− 4.05	− 1.33	− 2.04

4673	6109	2417	5984
− 3341	− 3007	− 1206	− 3713

Knock, knock. Who's there? Dogs. Dogs who?

Who's There?

Mixed review

Subtract.

58	67	76	85	94
- 37	- 26	- 15	- 34	- 73

42	78	92	65	63
- 26	- 59	- 28	- 49	- 25

790	456	720	954
- 85	- 109	- 212	- 728

58¢	72¢	44¢	28¢
- 9¢	- 8¢	- 19¢	- 12¢

Knock, Knock. Who's There? Cow. Cow Who?

FS111089 Grade 2 Subtraction